Make It Fit

Introduction to Tolerance Analysis for Mechanical Engineers

I0469710

Author: Jason Tynes
Edited: Carl Houk

Introduction to Tolerance Analysis

Table of Contents

Preface

Target Audience

This book is intended for students, recent graduates, and entry-level engineers. It can also be a good refresher for those who have been in industry and away from the design world long enough to forget the intricacies of tolerances analysis. Though it is intended to be read from cover to cover by students, recent graduates, and entry-level engineers, each person may benefit from this book differently.

Though some GD&T topics are periodically broached throughout the book (i.e. bonus tolerance, datum shift, geometric controls, etc.), this is not intended to be a GD&T intensive study. Ideally, the reader has access to the ASME Y14.5 standard, and time to review the rules governing the controls described therein. The intent of this book is to provide the reader with working knowledge on how to perform a tolerance analysis.

Students and Graduates

If you are still completing, or recently completed, your undergraduate or graduate engineering degree, this book will expose you to the area of tolerance analysis. Because you likely will not deal with the topic in great depth during your schooling, the content within these pages will give you a glimpse into the world of mechanical design, as well as give you a required skill for your first mechanical engineering job.

Entry-Level Design Engineers

As an entry level design engineer, the task of keeping a handle on part size and spacing and constraints can be daunting. In addition to thermal, structural, electrical, electromagnetic, and environmental requirements, there are three over-arching requirements levied on the mechanical designer. The system must:

1) Be able to go together.

2) Function as intended.

3) Be affordable (cost, weight, volume, etc.).

This book will show you how to determine if your design will fit or not; and also introduce you to ways to remedy tolerance- and assembly- related issues.

What is a Tolerance Analysis

Tolerance is the amount of variation associated with a feature or assembly constraint. All manufacturing and assembly processes have variations that impact a product. It may be the designer's intent to have a 0.500" diameter hole created one inch from either edge of a rectangular plate, but when the part is fabricated, the hole's size and location will rarely be perfect. The hole may be a little bigger or smaller, and it may be a little closer or farther from either edge. This allowable deviation from perfect is what is referred to as tolerance. In fact, the hole can have other variations besides size and location. The hole can have form tolerances, or be crooked – think of how straight a limp noodle is. The hole can also have orientation tolerances, or be non-orthogonal – think of the Leaning Tower of Pisa. The previous example illustrates manufacturing tolerances. Consider now the assembler who must install a screw into that hole. The hole should be slightly larger than the screw in order to allow the screw to install. When the assembler attempts to install the screw, the part with the hole in it could be free to move around as much as the screw/hole pairing allows. This is an example of assembly tolerances.

A tolerance analysis is simply a set of calculations that attempts to quantify the impact of manufacturing and assembly tolerances on a design's form, fit, and function. In a case such as a screw fitting through a clearance hole, a tolerance analysis can aid the designer in determining the proper hole size and/or tolerances associated with the hole. The tolerance analysis would be performed to determine the amount of clearance between the inside of the hole and the outermost surface of the screw. In the case of a loading a spring, a tolerance analysis would be an integral tool used in determining spring forces. The analysis would be used to calculate the length change of the spring. Though this book focuses on length-related tolerances, from the previous example, you can see how a tolerance analysis could easily be applicable to more than just variations in length.

Why should I perform a tolerance analysis

Because tolerance analyses are quantitative evaluations of a design, they reduce or eliminate speculation surrounding the design's adequacy. In other words, it removes some of the guesswork. Many inferior design options can be eliminated by performing a tolerance analysis.

When should I perform a tolerance analysis

Like most other analysis types, tolerance analyses are ideally performed in a product's design cycle, before parts have been manufactured. There are, however, times when a tolerance analysis is performed to validate a design change or identify a problem. In any event, the analysis should be performed when a quantifiable solution is desired. Tolerance analyses are most often performed only after enough design detail exists to justify the time spent in creating the analysis. It does not usually make much sense to perform an analysis on a design concept that is still in flux.

What tool(s) do I use to perform a tolerance analysis

The nature of tolerance analyses allows for the designer to utilize a wide variety of tools. Keeping in mind that a tool only makes a job easier and/or faster, the selection of the tool used to perform the tolerance is important, but certainly not more important than understanding the principles that govern the proper performance of the analysis. In theory, even the most complex, three-dimensional tolerance analysis can be performed on the back of a napkin with an old-fashioned pencil (eraser optional). However, there are tools that can make the task much easier and/or faster.

In all practicality, a spreadsheet is usually adequate for simple one- or two- dimensional analyses. Whether you use a company provided, commercially available, or personally developed spreadsheet, the underlying tool is similar. A spreadsheet is a tool where the user conceptualizes and builds the tolerance model with little assistance from the software package. This book focuses on simple analyses and, as a result, we will be using such a spreadsheet. We will be using QuickTol's QTA tool. It provides a format that is specific to tolerance analyses, is simple to use, and provides analysis outputs that are useful for our purposes.

Beyond one- and two-dimensional problems, however, a spreadsheet may become inadequate and cumbersome. There are commercially available tools that can assist the designer in building and performing a three-dimensional tolerance analysis with varying degrees of accuracy and ease of use. These tools are very powerful and require a non trivial amount of time to become competent with their use. Here is a short list of three-dimensional tools you may come across:

CETol: Developed by Sigmetrix, this is a tool that uses partial differential equations to develop sensitivity models and determine user defined measurements.

VSA: Developed by Siemens, this tool uses a Monte Carlo (random number generator) engine to create thousands of realistic simulations and determine user defined measurements.

3DCS: Developed by a company with the same name, this tool also utilizes a Monte Carlo engine to create simulations and determine user defined measurements.

Procedure

This procedure is intended to be a guide for people who are new to tolerance analyses. If followed in sequential order, it will aid in preventing the reader from being lost in their own analysis creation. While some analyses are significantly more complex than others, they all, generally, can be successfully performed using this procedure. If you follow these steps, even the complex analyses can be properly performed by hand when an expensive tolerance analysis software package is not available.

Each step in the procedure builds on the previous step. For example, in order to determine what actual requirement exists for a particular analysis, a gap must be identified. A gap is the space between the two features that represent the area of concern for a tolerance analysis (See Figure 6). It is usually a valueless exercise to create a tolerance loop when a failure criterion remains unknown. Here is an overview of the steps:

1. Identify the Potential Problem

2. Identify the Gap

3. Define the Requirement

4. Create a Tolerance Loop

5. Create a Table for Calculations

6. Research Tolerances and Populate Table

7. Calculate the Gap

1) Identify Potential Problem

Identifying the problem is the start of any tolerance analysis process. Here, we ask a pair of questions: What problem am I trying to solve, and how do I know when there is a problem? In most cases, it is very simple to see. For example, Figure 1 illustrates an obvious interference between a screw and mounting hole. A screw going into a hole presents a problem when the outermost surface of the screw attempts to go through the edge of the hole.

When this scenario is presented, it is obvious to see that the screw could have trouble being properly installed.

Figure 1 Obvious Problem with Screw/Hole Pairing

Other cases may not be as easy to identify, but identify, we must! We cannot move on to the next step and identify the gap unless we know the potential problem we are trying to quantify. Skipping this step can easily put you on the path of calculating the wrong gap. Here are a couple examples to help you hone your skills:

Figure 2 shows a cross section through a DC motor within a housing. The motor is attached to the housing with screws and the motor shaft is permitted to protrude through the front of the housing, and rotate unimpeded. At least, that's how it is supposed to work. We have identified three areas (A, B, and C) that may be problems. Areas A and B indicate areas where the motor may contact the housing walls. If one of these areas were shown to have a potential of contacting during assembly, it may prevent the motor from installing altogether. Area C is a little different. Take a minute or two to consider why area C is important.

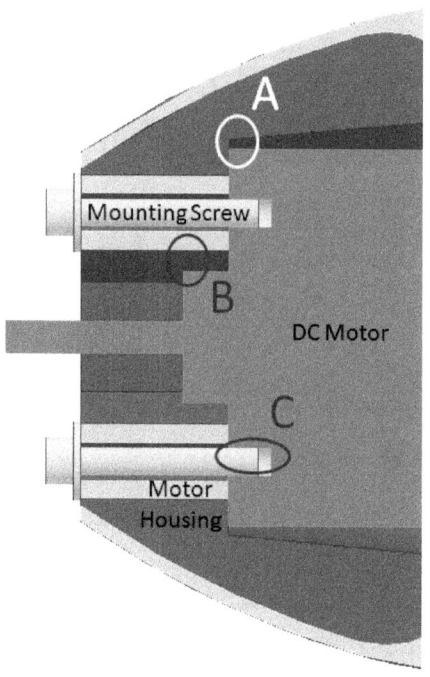

Figure 2 DC Motor within Housing

Area C represents the screw engagement area. Presumably, the screws are there to keep the motor in place. However, the screws require a minimum number of engaged threads in order to adequately perform their function. If the engagement is insufficient, the joint may fail (the motor falls off) during operation. If there is too much engagement (the screws are too long), the screw may interfere with the operation of the motor or bottom out in the hole. Either scenario (screw too long or screw too short) could cause undesirable conditions. Let's try another example.

You were tasked with evaluating a door and latch assembly that interfaces to a door frame, as shown in Figure 3. What could go wrong with the door that would indicate a functionality problem?

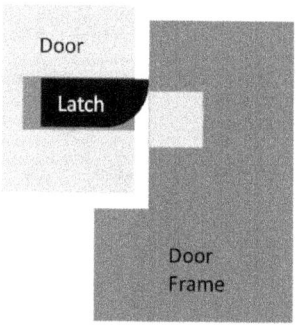

Figure 3 Door and Latch Assembly

While a list of remote problems could go on forever, there are only a few viable things that could go wrong with the door:

1. The latch may not prevent the door from re-opening, once closed.
2. The door may swing through and beyond the closed position (i.e. the door jam does not stop the door).
3. The door may not close all the way (i.e. the door binds up on the jam).

Because Figure 3 only shows the latch area, we will focus our discussion here on the potential problems that relate to the latch – the first item on the list. The only way that the door would be prevented from remaining closed is if the latch were not within the slot of the door frame. (This assumes, of course, that the door, latch, and frame remain intact!). So the next logical question is: How do you know if the latch will go into the slot of the frame? That is the next step: Gap Identification.

2) Identify the Gap

Identifying the gap is critical because it focuses the designer's efforts on evaluating the real potential problem area. The numerical value of the gap is what will be quantified.

In determining if the latch will go into the slot within the door jam, it is imperative to take a close look at how the door assembly is used in normal operation. The door and latch assembly is swung until the door panel contacts the door frame. The position of the latch, when the door contacts the frame must be conducive to allowing the latch to fall within the slot.

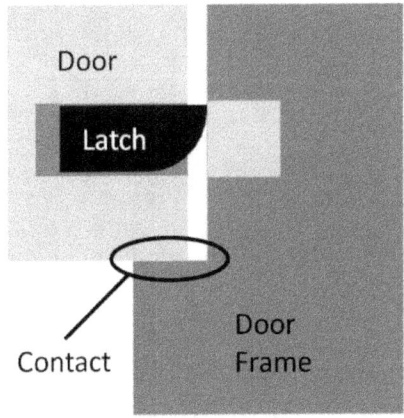

Figure 4 Door and Latch Assembly in Contact with Frame

Figure 4 shows the door swung until it makes contact with the door frame. We can now see quite clearly that the latch is able to engage the slot only if the top edge of the latch has moved below the top edge of the slot. Therefore, this area is what we identify as the gap as shown in Figure 5.

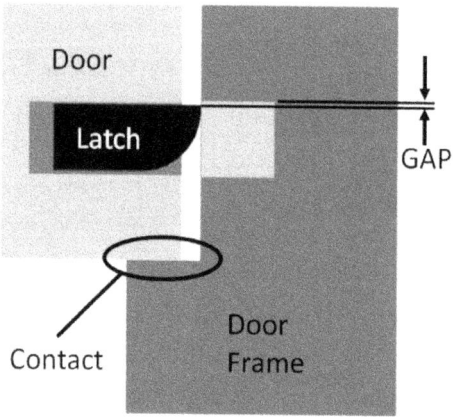

Figure 5 Door with Gap Identified

3) Define the Requirement

The requirement is the value (or range of values) that represents the cutoff between an acceptable and unacceptable gap or interference (lack of a gap), in some cases. It is answering the question of: how do I know when the gap is good or bad. In the simple screw and hole example (see Figure 6), the requirement is that the gap must be greater than or equal to zero in order to be deemed acceptable. The determination to include or exclude zero as an acceptable gap value is something that must be done using sound engineering judgment. If the value is indeed zero in the real world, is that really acceptable? You must answer that for yourself and it is dependent on the part/assembly being analyzed.

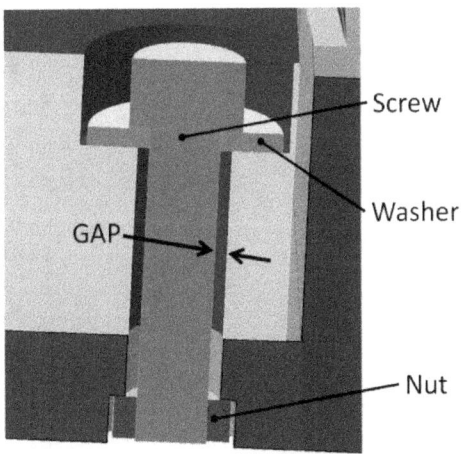

Figure 6 Sample Screw Gap

In some instances, a range of gaps is acceptable. Consider the example where a gasket is being compressed. If the gasket is not compressed enough, it could leak. That means that the compression must be some minimum amount. Conversely, if the gasket is compressed too much, it may tear, allowing for it to permit leakage yet again. This indicates the need for a maximum compression requirement as well. Whereas most analyses only have one-sided requirements (minimum clearance of zero, etc.), this type of requirement is often referred to as a two-sided requirement. Too far to either side, and there could be trouble. We must find the sweet spot.

4) Create a Tolerance Loop

A tolerance loop is the feature and assembly constraint path that locates one side of a gap relative to another. It is usually pictorially represented in order to quickly convey which tolerances play into the analysis. The tolerance loop, ideally, begins at one side of the gap and ends with the other, by systematically stepping from a feature of a part to another, through joints, and on to features of the another part. The path traversed either comes directly from the dimensioning scheme of an engineering drawing or subsequently defines the dimensioning scheme on an engineering drawing. We will revisit this concept in more detail within this section. See Figure 7 below. Notice that the tail of V1 (Vector #1) coincides with the bottom side of the gap. The tail of V2 coincides with the head of V1. The vectors are placed head-to-tail. You will likely recall that this is identical to the way most of us learned to perform vector addition. Put all the vectors head-to-tail, and the resulting vector (the vector that goes from the tail of the first vector to the head of the last vector) is the gap you were looking for. This is simple vector addition.

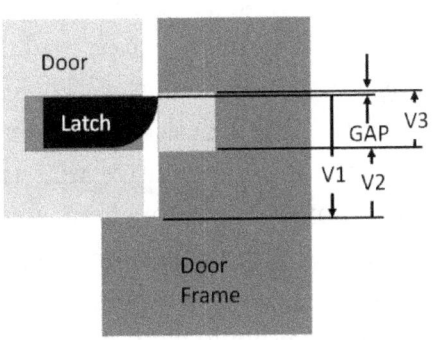

Figure 7 Sample Tolerance Loop

For correctness, it is imperative that the path remain unbroken. A broken path (imagine if the tail of V3 were not coincident with the head of V2) does two things: it violates the rules of vector addition, and it overlooks the impact of a variable within the analysis. This results in an overly-rosy representation of the tolerance impact on a design. It is also important that the analysis does not repeat the same part or feature. A loop that backtracks on itself is a clear indication that the wrong constraint path has been chosen.

Choose a Starting Place
It is usually best to start at one side of the gap. This enables the designer to proceed in a single direction, while maintaining a single and consistent sign convention. Sign conventions will be addressed in the next section.

Choose a Sign Convention
Because most engineers have it engrained in their brains that right and up is "positive", it is usually helpful to maintain that sign convention in tolerance loops. For this reason, it is recommended to start on the left or bottom side of a gap. In starting on the left or bottom, positive solutions translate to "gaps", while negative solutions translate to "interferences". There is no rule however that states that a designer must start on any particular side, but it is imperative that the positive direction remain consistent throughout any particular analysis. If you decide to make left (or down) positive, a simple arrow pointing left (or down) with a plus-sign on your tolerance loop should convey to any reader that left (or down) is positive.

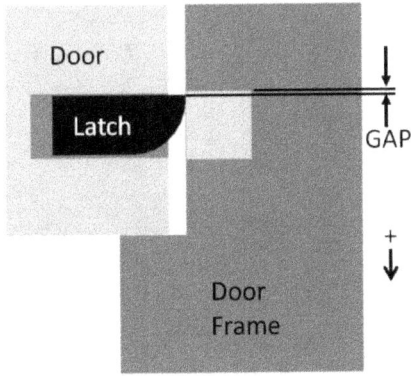

Figure 8 Down is positive example

Path Definition

The relationship between the tolerance analysis and drawing cannot be negated. The decisions made during the creation of one will implications for the other. One area that the two must agree on is the dimensioning scheme. The path utilized in the tolerance analysis comes directly from the drawing. Let's consider the drawing for the door jam of Figure 7, as shown in Figure 9.

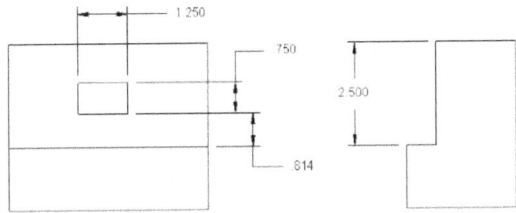

Figure 9 Actual Drawing Except for Door Jam

Closely inspecting the drawing, we notice that the drawing and tolerance loop are consistent. That is, the distance from the top edge of the slot to the surface that contacts the door is comprised of two dimensions (.814 and .750). These two dimensions correspond to vectors

V2 and V3, respectively. Take a look at Figure 10 . It shows a different dimensioning scheme from Figure 9, as indicated by the 1.564 dimension.

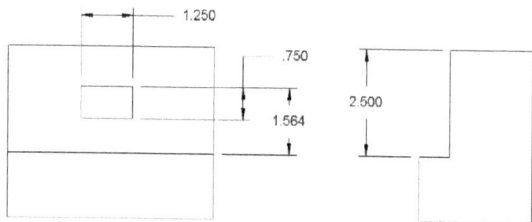

Figure 10 Hypothetical Drawing Excerpt for Door Jam

Yes, the nominal location of the top side of the latch area is still 1.564 from the stop, but the path shown in the tolerance loop does not agree. Had the door jam drawing actually looked this way, vectors V2 and V3 would be incorrectly depicted in the tolerance loop. Though the numerical impact of this particular inconsistency may be minor in this case, the point here is clear: the tolerance loop cannot be created in isolation with respect to the detailed drawing. Unless the drawing is nonexistent at the time of tolerance analysis creation, the detailed drawing must drive the tolerance loop.

In an ideal product design environment, the tolerance analysis is created in parallel with the 3D model. The engineer comes up with an idea, it is modeled, and the tolerance analysis (along with other first-order analyses) is created alongside, to give credence to the design concept. Typically, a detailed drawing does not exist at this point in the design cycle. The tolerance analysis drives the dimensioning scheme and defines the tolerance values. In order to prevent tolerance buildups from negatively impacting the design, it is common to assume a dimensioning scheme that provides the shortest tolerance loop possible for the most critical gaps. For example, let us assume that the gap show in Figure 7 is the most critical gap for the door assembly design. Which dimensioning scheme would provide the shortest possible tolerance loop, Figure 9 or Figure 10? The shortest tolerance loop is the one that requires the least number of vectors to navigate from the starting side of the gap to the end. Therefore, Figure 10 would provide the shortest possible loop because it combines vectors V2 and V3 into a single vector. If this gap were critical, we would likely want to set the dimensioning scheme up in this manner.

Understand the Constraints

Constraints are the physical restrictions within an assembly or part that govern or limit the degrees of freedom of a part or feature. There are two common constraint types: Assembly and Feature. Most of the job of performing an accurate tolerance analysis is rooted in being able to visualize how the gap is impacted by each constraint type.

Feature Constraints

Feature constraints are possibly the easiest to understand and categorize. These are constraints found on the face of detailed drawings. These are the plus-or-minus, positional, perpendicularity, profile, and concentricity tolerances, just to name a few. They control and limit the feature-to-feature relationship within a part. Thousands of trees are chopped down each year to produce the volumes of information related to the documentation and interpretation of such constraints. I won't add to the chatter, because in all likelihood, you will need to review the ASME Y14.5 standard for clarification at some point in your career. And truthfully, the rules and interpretation of the rules can change every decade.

Assembly Constraints
Assembly constraints pick up where feature constraints left off. These are the constraints you will not find on a drawing (in the vast majority of cases). Also unlike feature constraints, there is not a library full of information about the rules governing assembly constraints. They are the constraints that dictate the location and orientation of a feature(s) of one part to a feature(s) of another part. Although assembly constraints still control a feature-to-feature relationship, the features are on different parts. Therefore, they control the feature-to-feature relationship between parts. The most common among this group is assembly slop/float. A perfectly located pin (within part A), going into a perfectly located hole (within part B), can still have float between the two, depending on the size difference of the features. With assembly constraints, we have to do *some* thinking (not usually a whole lot!) about what happens at the joint between parts. Figure 11 below identifies the assembly slop as V3.

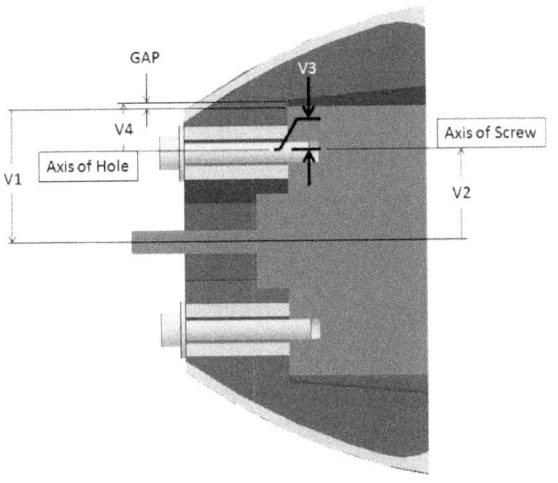

Figure 11 Assembly slop

Though it can be depicted in many different ways, the numerical representation is should be consistent, if properly accounted for. In this figure, the assembly slow can be visualized as the distance between the axis of the screw and the axis of the corresponding hole. Nominally, that distance is zero, but depending on the size difference between the screw and hole, it can be a non-zero value. In other words, the screw can be pushed up against any edge of the hole. Where it actually lies within the hole, in the real world, can depend on any number of manufacturing conditions. For example, an assembly aid could minimize the amount of float at the joint, or unmitigated gravitational forces could bias every assembly in a single direction to the maximum allowable extent.

Though much less common than assembly slop, another type of assembly constraint is a shimming operation. Used in the assembly sequence, this operation can assure a certain sized gap between parts before their relative locations are finalized. This shimming operation would control the feature-to-feature relation between the two parts. Typically, these types of constraints can only be found in assembly instructions, and occasionally on an assembly drawing.

Other Constraints
There are special cases where other types of constraints can come into play that do not cleanly fall into one category or another. One such constraint is an envelope-controlled dimension for an assembly. Companies work together to develop a single product every day. Often times detailed part information is not shared between the two companies, but rather an envelope controlling document of some kind. This type of documentation is used to communicate design volume, interface details, and other very useful information. For now, we will focus on the volume and interface details. Though the envelope defined in the document may appear to be a single part, it could very well contain multiple piece parts. Detailed information on the piece parts within the volume is not always necessary in performing a conservative tolerance analysis. Let's say company A is designing a widget that is physically located near the volume where company B's sprocket is being designed to fit within. If company B is designing their sprocket to fit within a documented and pre-allocated envelope, company A only needs to assure that the widget does not invade company B's volume. In effect, the tolerance analysis that verifies proper fit between the widget and sprocket (as view by company A) can be simplified by

treating the sprocket as a single, lumped part, as defined by the envelope document.

5) Draw Vectors and Label Constraints

Each of the constraints of a tolerance analysis should be shown as a vector on an image that is representative of the hardware being analyzed. The vectors should guide the reader from one feature to another, following the constraint path defined by the hardware. This allows the reader to visually interpret which value (nominal and tolerance) is being represented by the particular vector. As stated previously, each vector should correspond to a particular dimension on a drawing (or model). Comparing the tolerance loop to the available documentation should provide the most accurate path through the parts as possible. Each vector should also be named, or otherwise identified, for cross-reference to the calculation.

If we were to consider the vectors involved in the analysis of Figure 7, the vectors would be named something like this:

V1: **From** top edge of latch pocket within door **to** door frame interface

V2: **From** door frame interface **to** front side of slot in door frame for latch

V3: **From** front side of slot in door frame for latch **to** back side of slot in door frame for latch (aka: width of slot)

This path takes us from the top edge of the latch pocket within the door to the back side of the slot in the door frame for the latch. In other words, "the gap".

Using the following figure and drawing excerpts as references, practice drawing tolerance loops for the three gaps shown in Figure 12.

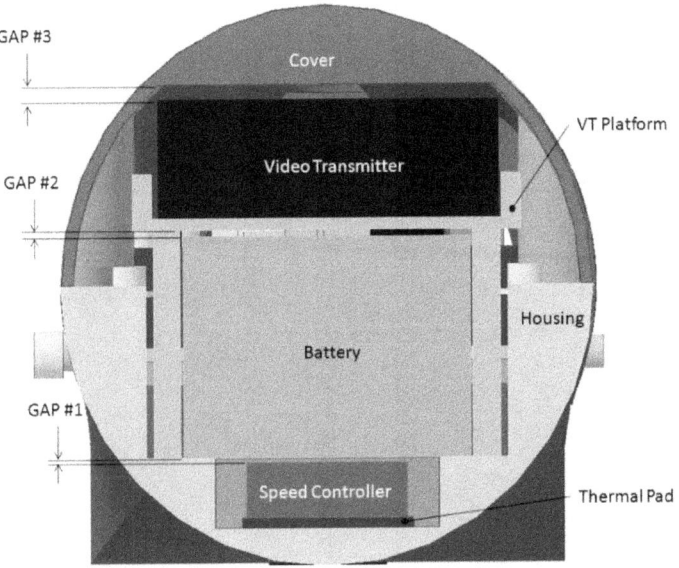

Figure 12 Practice tolerance loops

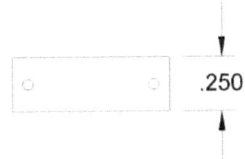

Figure 13 Speed Controller

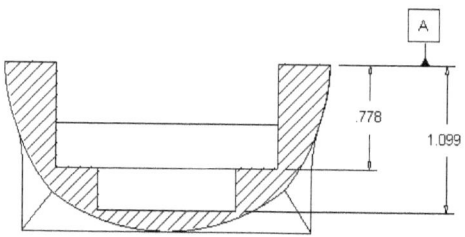

Figure 14 Housing

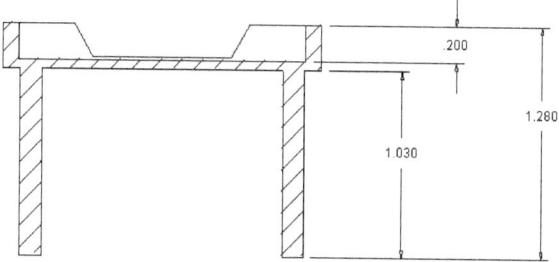

Figure 15 VR Platform

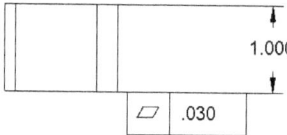

Figure 16 Battery

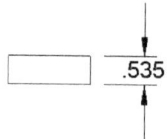

Figure 17 Video Transmitter

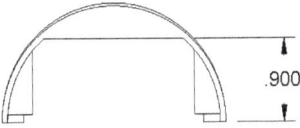

Figure 18 Cover

Gap #1 represents the space between the top of the video transmitter and underside of the cover. Assuming the thermal pad is .050 thick, draw the tolerance loop on the image below.

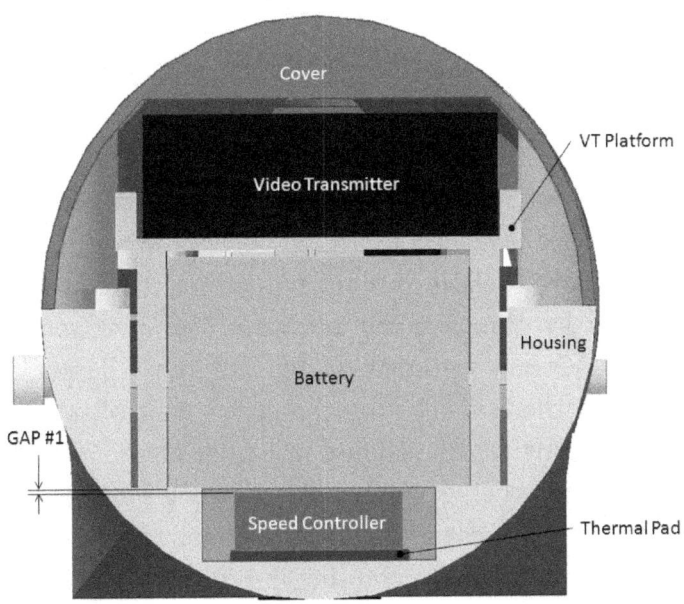

Figure 19 Gap #1 Example

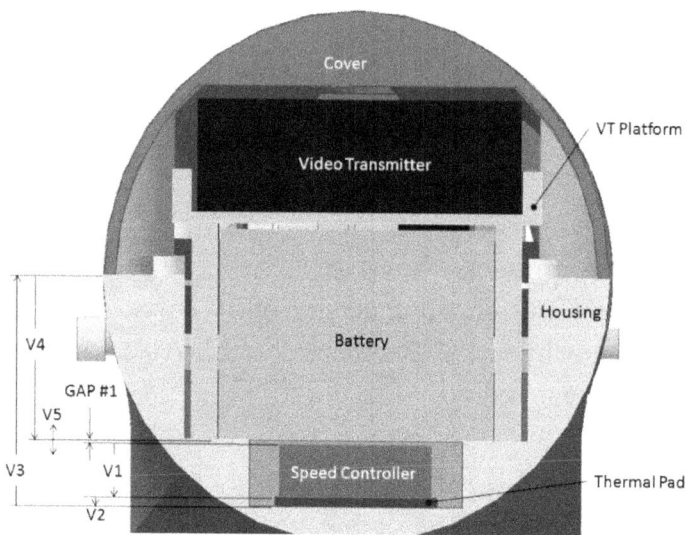

Figure 20 Gap #1 Tolerance Loop Solution

Starting out on the bottom side of the gap (for convenience), we loop around to the opposite side of the gap. This is a relatively simple loop, but there are two noteworthy things going on here. The first is that vectors V3 and V4 combine to determine the pocket depth in the housing (for the speed controller). The dimensioning scheme shown in Figure 14 prevents the vectors from being combined to a single dimension. Second, we have added a new symbol to the image. Vector V5 has arrows pointing in both directions. This is common when identifying tolerances that are nominally zero, but can be either positive or negative values in the real world. In this particular case, the symbol is representing the flatness tolerance on the battery. Because the battery mounts to the housing on only on the edges of the battery, there could be a droop on the bottom surface of the battery that impacts the gap. As a matter of fact, this bottom surface could be concaved down or up, which would reduce or increase the gap, respectively.

Gap #2 represents the space between the top side of the battery and underside of the VT Platform. Draw the tolerance loop on the image below.

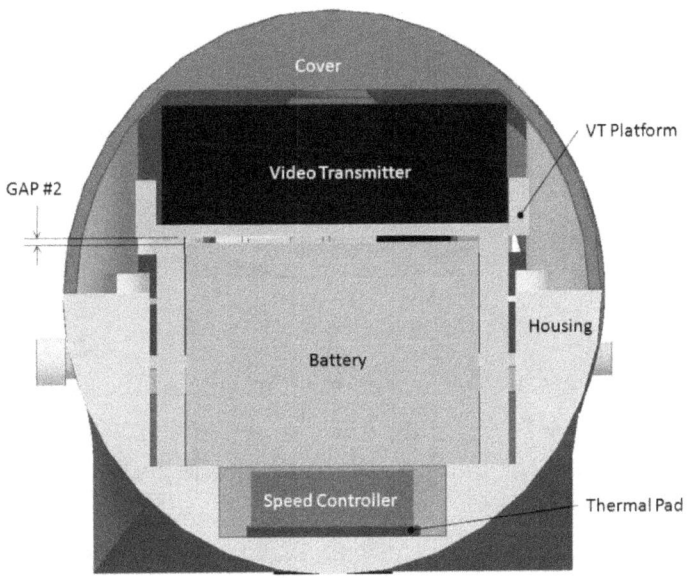

Figure 21 Gap #2 Example

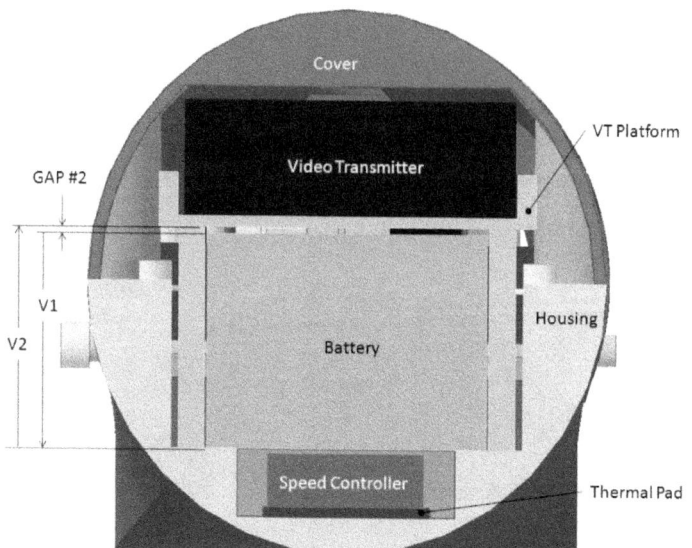

Figure 22 Gap #2 Tolerance Loop Solution

Again, we start out at the bottom side of the gap and loop around to the opposite side. Notice that we are not considering the flatness tolerance on the bottom side of the battery. That is because the flatness tolerance can do nothing but increase this gap. A battery that is concave up still must meet the 1.000 dimension on the battery drawing. We have made a conscious decision to exclude the flatness tolerance. The take away from this example is that you must consider the potential impact or a tolerance before including or excluding it. After a bit of practice, it becomes much easier to identify.

Gap #3 represents the space between the top side of the video transmitter and underside of the cover. Draw the tolerance loop on the image below.

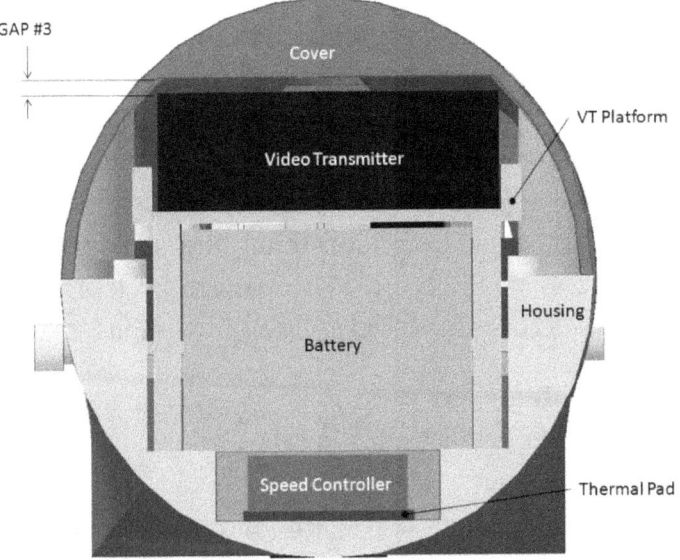

Figure 23 Gap #3 Example

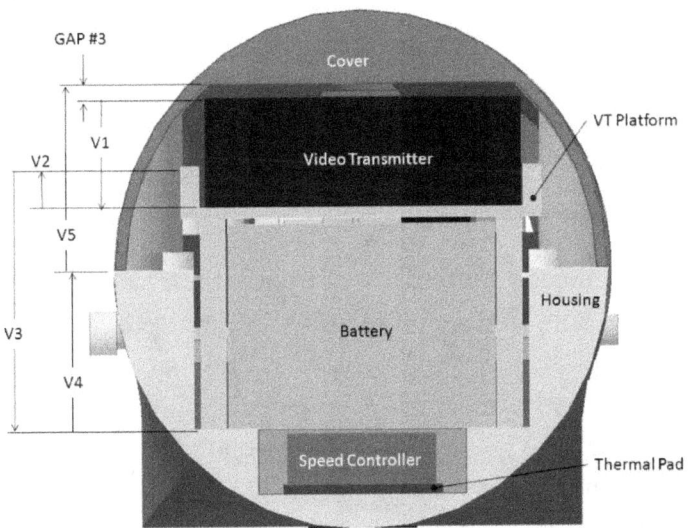

Figure 24 Gap #3 Tolerance Loop Solution

6) Tolerance Research and Calculation

Researching and calculating tolerances is the step where nominal and tolerance values are researched within drawings and models and used to calculate any applicable tolerances. On occasion, documentation is lacking for a particular constraint and an assumption must be made. Early on in the design cycle, the vast majority of the values used will be assumptions.

Researching

Pouring over detailed drawings to find the nominal and tolerances values, the designer should be combing through dimensions that correspond to each vector of the tolerance loop. Though this is listed as a step in the process following the creation of the tolerance loop, it is best done in parallel. Be aware of any inconsistencies between the tolerance loop and the dimensioning scheme of drawings. When there is a difference between the tolerance loop and dimensioning scheme, the most appropriate thing to do is to correct the tolerance loop to reflect the documentation.

As technology continues to advance, there will likely be fewer and fewer drawings required to produce products. It is very probable that drawings will go the way of cassette tapes and rotary phones. Though drawings are going away, dimensions are still required to communicate design intent. When drawings go away, the need to perform a tolerance analysis has not been eliminated; the dimensions and dimensioning scheme has just moved from the drawing to the model. It will be necessary for the tolerance analyst to be able to interrogate the 3D model to establish the relationships/constraints for features. Essentially, when the drawing goes away, we get the information we need from the model and the tolerance analysis process continues on as usual.

Calculating

Each dimension (i.e. vector or constraint) that has been researched should be recorded (typically in a spreadsheet or other specialized tolerance analysis tool) in a manner that is consistent with the sign convention and tolerance loop.

Nominal Values

If you are using a right-is-positive (or up-is-positive) sign convention and a particular vector in the tolerance loop is pointing to the left (or down), the nominal value associated with that constraint should be negative. The magnitude of that nominal value should match that of the drawing. In the vast majority of cases, this should be a simple copy/paste effort.

Creating a simple table in a spreadsheet program is usually the fastest and most efficient way of documenting a simple tolerance analysis. With that in mind, here is an example of the type of table that can be created for the gap shown in Figure 7.

Introduction to Tolerance Analysis

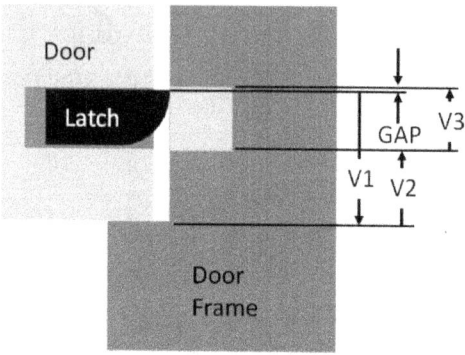

Figure 25 Tolerance Loop

Vector Name	Description	Mean	Tolerance
V1	From top edge of latch pocket within door to door frame interface	-1.05	
V2	From door frame interface to front side of slot in door frame for latch	0.65	
V3	From front side of slot in door frame for latch to back side of slot in door frame for latch (aka: width of slot)	0.65	

Table 1 Example Calculation with Nominal Values

Tolerances

Calculating the tolerances associated with a particular dimension can get tricky. It is important to rely on the tolerance loop here. Depending on the path, the tolerance may need to be expressed as a total tolerance value, half the tolerance value, or in some cases, doubled the tolerance value. The key is to understand how the variation allowed by the specified tolerance impacts the gap in question. In the case of a hole axis being located from the edge of a panel, it could be located using a plus-or-minus dimensioning scheme or a geometric positional tolerance. Each will need to be treated differently.

Let's determine the tolerance allowed on the .175 dimensions in the figures below. The ±.010 tolerance from Figure 27 can be used as is, but the geometric positional tolerance in Figure 26 needs to be adjusted in order to accurately represent the constraint of concern. This is because the geometric positional tolerance control utilizes a circular (technically cylindrical) tolerance zone with a diameter equivalent to the tolerance value (.010) to contain the axis of the hole. The axis of the hole is therefore allowed to reside anywhere within half (.005) of that circular zone. It is allowed to be one radius (diameter divided by 2) up or down. The location of the hole axis is ±.005.

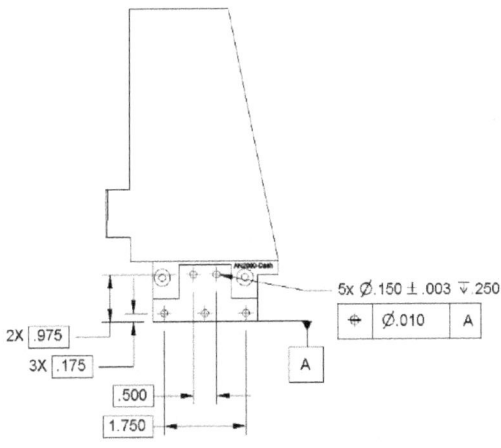

Figure 26 Hole Dimensioned with Geometric Tolerancing

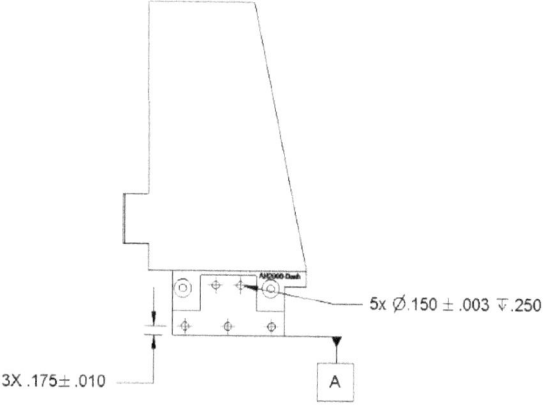

5x ⌀.150 ± .003 ▽.250

3X .175±.010

Figure 27 Hole Dimensioned with Plus-or-Minus Tolerancing

Using the plus-or-minus dimensioning scheme from the previous example, if we were attempting to calculate the virtual size of the hole (the size the hole could appear to be from a top view), the tolerance would need to be doubled. In this case, the virtual size is the diameter (technically the smallest diameter for a hole, or .147) of the hole minus the diametric tolerance associated with its orientation. Because the plus-or-minus dimensioning scheme could be likened to a radial tolerance, it must be multiplied by two to calculate the diametric tolerance.

Expressing the tolerance is as simple as recording it in the spreadsheet program. Adding the tolerances to Table 1, we can create Table 2.

CAUTION: This example is one-dimensional. Converting a plus-or-minus dimension to an equivalent cylindrical geometric tolerance zone is not as simple as multiplying by two when performing a two-dimensional analysis. Plus-or-minus tolerancing produces rectangular tolerance zones, while positional tolerances on holes are typically circular tolerance zones. See Figure 28. Though both the rectangular and circular tolerance zones allow for the axis of the hole to reside within zone A, the rectangular tolerance allows for the axis to reside in the corners (Zone B). The circular zone prohibits the hole from being in Zone B.

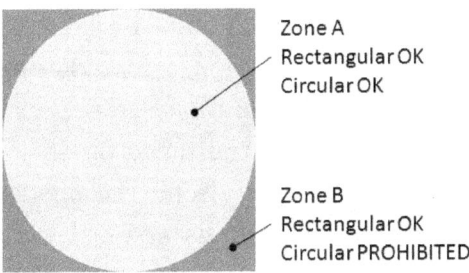

Zone A
Rectangular OK
Circular OK

Zone B
Rectangular OK
Circular PROHIBITED

Figure 28 Circular vs. Rectangular Tolerance Zones

Vector Name	Description	Mean	Tolerance
V1	From top edge of latch pocket within door to door frame interface	-1.05	0.1
V2	From door frame interface to front side of slot in door frame for latch	0.65	0.075
V3	From front of slot in door frame for latch to back side of slot in door frame for latch (aka: width of slot)	0.65	0.025

Table 2 Example Calculation with Tolerance Values

Recording the Tolerance

The decision to display the tolerance as a plus-or-minus value or a diametric value is a matter of personal preference. However, as a word of caution, most engineers display all tolerances as a nominal value and a plus or minus tolerance. The nominal value is the mean between the minimum and maximum limits on a particular dimension. This is referred to as nominal and bilateral. It is usually convenient and well understood to communicate dimensions in this fashion. Unilateral (.125 +0/-.005) or unequal bilateral (.125 +.002/-.005) dimensions should be adjusted accordingly (.1225 +/- .0025 and .1235 +/- .0035 respectively). As a word of caution, depending on how the part is modeled, this approach may produce a mean shift, preventing the nominal condition of the tolerance analysis from matching identically with the model. Had the part been modeled using a .125 dimension, we would not be using the same, .125, dimension in the tolerance analysis. We have successfully covered the range on the value, but shifted the analysis away from the model's value. Though the model and analysis nominal values would match, neither would be technically incorrect. No matter your preference, it is absolutely imperative that the values shown in your calculations agree with the tolerance loop and the other values within the analysis. Mixing diameters and radii should be done with the utmost care.

Assumptions

It is not uncommon to come across features that are poorly or inadequately controlled. A feature may be unconstrained, under-constrained, or even over-constrained (with conflicting values). It can also be vaguely constrained with an ambiguous dimensioning scheme. When these potential issues are encountered, an assumption must be made. It is best to make a conservative assumption, if a conservative assumption can be made at all, and document that assumption. Use a placeholder value(s) in the interim and seek advice from more senior engineers or manufacturing representatives for suitable values. The worst thing that can be done is to ignore the oversight, assign a zero for that tolerance, and move on. Bringing unknowns to the table for discussion and resolution is the best way to deal with such an issue, provided you do not already have enough experience to assign an appropriate value.

7) Solution

The solution is the answer that is calculated in the spreadsheet or tool. Typically, the solution is expressed as a nominal and bilateral tolerance. The values correspond to the nominal gap and total tolerance accumulation for the gap. For one-dimensional analyses, the solution is typically a simple summation of all tolerances. The solution for the simple example we have been looking can be determined by summing the nominal and tolerance values, respectively.

Vector Name	Description	Mean	Tolerance
V1	From top edge of latch pocket within door to door frame interface	-1.05	0.1
V2	From door frame interface to front side of slot in door frame for latch	0.65	0.075
V3	From frontof slot in door frame for latch to back side of slot in door frame for latch (aka: width of slot)	0.65	0.025
	Gap / Solution	0.25	0.2

Table 3 Example Calculation with Gap/Solution

Table 3 shows that the gap is 0.250 ± 0.200. This means that the gap can range anywhere between 0.050 and 0.450. Comparing the solution to the requirement, we see that the gap is always greater than zero. For this particular case, the analysis confirms that the design meets this specific requirement. Had the requirement been 0.100 minimum, the design would not necessarily meet the requirements. At that point we may consider evaluating the analysis on a statistical basis.

Bringing it all together

Let's revisit a previous analysis to bring it all together. Bringing back the gap from Figure 19, we recall that we already developed the tolerance loop as shown in Figure 20. Here it is again as a refresher.

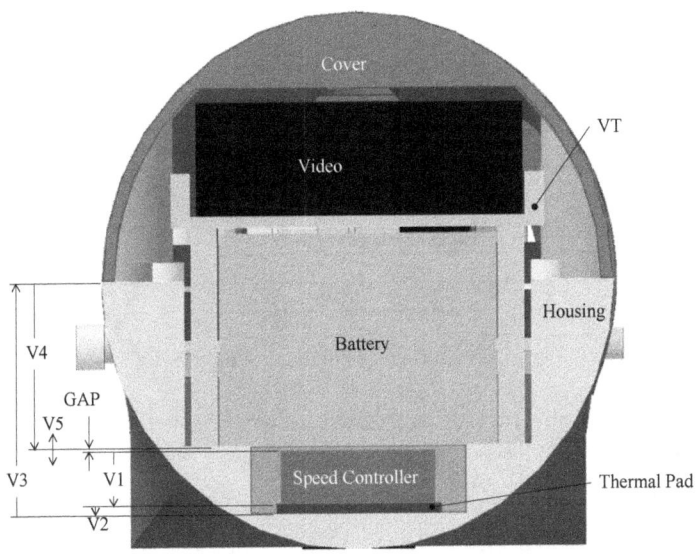

Figure 29 Gap #1 Tolerance Loop

The potential problem we are trying to avoid here is that the speed controller generates a lot of heat. We would like to avoid having that heat transferred to the battery. The requirement, therefore, is to have a gap that is greater than .005". Let us assume for the time being that the default tolerance for each part is ±.005".

Using the drawing information from earlier, we can fill out a spreadsheet with the analysis information.

	A	B	C	D
1	Gap Description	Speed controller to battery gap		
2				
3	Vector Name	Description	Mean	Tolerance
4	V1	Thickness of speed controller	-0.25	0.005
5	V2	Thickness of thermal pad	-0.05	0.005
6	V3	From bottom of speed controller pocket within housing to datum A of housing	1.099	0.005
7	V4	From datum A of housing to bottom of battery pocket within housing	-0.778	0.005
8	V5	Flatness of battery	0	0.03
9				
10		Gap	0.021	0.05

Figure 30 Gap #1 Tolerance Analysis

Notice that the flatness tolerance from Figure 16 is represented here as a plus-or-minus value. This is because there could be a .030" step on the bottom of the battery. This step could be .030 up or .030 down, hence the plus or minus .030 tolerance.

The gap is calculated simply by summing the nominal values together and doing the same for the tolerances. We end up with a solution, or gap, of .0210 ± .0500". Consequently, we could have a gap as large as .071", or interference as large as .029". Because the requirement is a minimum of .005" gap, there is a potential to violate the requirement.

Statistical Considerations

Depending on a variety of circumstances, the solution may be evaluated on a worst case or statistical basis. Each evaluation method has its benefits and its drawbacks. A worst case analysis regards simply the minimum and/or maximum solution, while a statistical analysis is geared more toward determining the probability of meeting design intent. Simply put, worst case represents what <u>could</u> happen, while statistical represents what is <u>most likely</u> to happen.

Worst Case vs. Statistical

Worst case analyses are typically preferred in situations where an excessive accumulation of tolerance could result in potential loss of life or high value asset(s). Within some industries however, worst case analyses are considered overly-conservative when the number of variables in an analysis exceeds some finite number (usually 4), regardless of the risk associated with a failure.

Statistical analyses are widely preferred in low risk situations where an excessive accumulation of tolerances could result in a failed build, but no significant losses. A statistical analysis reveals the probability of encountering an undesirable condition in any given build. These probability values can be used to support financial models associated with the manufacture and sale of a product. Because statistical analyses describe what is most likely to happen, they tend to represent reality much better than a worse case analysis, provided the inputs also represent reality. We rely on probabilistic outcomes every day. When we drive to work every morning, a worst case scenario is that we could be seriously injured in a car accident. However, we will probably arrive to work without incident. The worst case scenario does not prevent capable drivers from using automobiles. Similarly, a worst case tolerance analysis should not keep good designs from being produced when the probability of producing good product is acceptable to the producer and customer.

Because of the benefits of each type of analysis, it is common to use both methods in evaluating a solution. In many specialized tolerance analysis tools, the software provides results based on both methods.

Statistical Measures

There are quite a few measures used to describe statistical analyses. Some speak to the amount of variation expected, while others address the probability of encountering a problem. Variation measures include:

Standard Deviation – Amount of variation about the mean or expected value. A low standard deviation indicates a low amount of variation.

RSS – Root Sum Squared. Rough approximation of amount of variation expected. See the RSS area for more information.

MRSS – Variation of RSS. Adds some conservatism, but still provides a rough approximation of amount of variation expected.

Measures that address the probability of encountering a problem are admittedly more useful to any decision-making process. Ultimately, though, knowing the probability of encountering a problem does not mean a whole lot if there is no context to process that information. Simply put, what is considered "good enough"? If a design has an expected yield of 99.9%, is that good enough to move forward to production? Here are some common measures:

Yield – The probability of encountering a favorable condition.

PNC (Probability of Noncompliance) – The probability of encountering a condition that does not conform to the requirement. Equivalent to 100% - yield.

Sigma (or Sigma Level) – Describes the number of standard deviations between the mean and requirement.

CpK – Scalar representation of a process' ability to produce outputs within a specified range.

Tolerance Distributions
Because statistical analyses are such powerful business decision tools, it is imperative that the analysis outputs reasonably accurate probabilities. A large driver of inaccuracies of statistical analyses is the tolerance distribution used for individual variables. In short, how good is your manufacturer or assembler at creating near-perfect features and assemblies? Are you just as likely to get a perfect feature as you are to get a feature at one tolerance extreme? Will the features most likely be closer to the nominal value than an extreme? This type

of information, going into a statistical analysis, has an impact that cannot be ignored.

Uniform

A uniform distribution represents the case where a feature or tolerance is equally likely to be anywhere between its limits. For example, a hole that is sized at 1.000 ± 0.500 is equally likely to be 0.500, 1.000, or 1.500. Uniform distributions are usually conservative assumptions in the absence of manufacturing data. However, most manufacturing processes do not produce uniformly distributed features. In fact, there may be cases where a manufacturer intentionally biases the size of a feature to one size extreme or another. In such a case, a uniform distribution would not necessarily be a conservative distribution.

Gaussian

A Gaussian distribution represents the case where a feature or tolerance is more likely to be closer to the nominal value (or some other, shifted mean value) than not. It is the classic normal or bell-curve. The vast majority of manufacturing processes produce features with Gaussian distributions. However, skewness, kurtosis, mean, and standard deviation are parameters that describe the size and shape of the Gaussian distribution and cannot be very accurately approximated without knowledge of the process used to create each feature. There is a popular idea circulating that tolerances can be assumed to be normally distributed with three-sigma accuracy. It is potentially dangerous to assume that all tolerances are normal distributions, because a normal distribution is a

Gaussian distribution with very specific parameter settings. These settings are not necessarily appropriate for all tolerances and manufacturing and assembly processes. An inaccurate and non-conservative assumption here could prevent the designer from identifying a potential problem early in the design phase.

Choosing a Distribution

It is my personal preference to ask a knowledgeable manufacturing representative about the distribution to consider for a particular feature. In the absence of a knowledgeable person, or statistically significant data points to create a more reasonable distribution, I prefer to use a uniform distribution. The uniform distribution may send up more false red flags, but I am almost certain to be made aware of all the problematic areas of a design with this approach.

Communicating Distribution with Manufacturers

Pretend for a moment that you are the technician tasked with cutting a piece of bar stock to the length of 20.000 ±0.500". This is easy enough to do, even at home, nearly blindfolded. Without more information, we could probably assume that most of the bars you cut will be closer to 20.000" than 20.500" or 19.500". We could *assume* that you would aim for the middle of the tolerance band: 20.000". However, we have not considered all of the factors involved in your decision on what size to aim for. In fact, you could be incentivized to aim at sizes other than 20.000".

Consider, for example, that you need to cut multiple parts from the same stock. You could easily aim for a lower value, say 19.510", if you know your machinery can easily hold tolerances within ±0.010". In doing this, you would be able to save nearly 1" of material with each part you make. In essence, you can create more acceptable parts from a single piece of raw material. You would effectively be targeting a value that is significantly lower than the mean, producing parts within a range of 19.510 ±0.010". Substantially different from the 20.000 ±0.500, yet still perfectly acceptable parts.

Now let's consider a different scenario. If the bar stock you were to be starting with were high-cost and extremely brittle, you would certainly not want to aim for the low side. In fact, you may be prone to aim for the high side, say 20.490" if you know your machinery can easily hold tolerances within ±0.010". The impact of choosing the high side to aim for allows for you to have enough material to rework a part in the event that a crack develops. In effect, the only time a part is produced with a size other than 20.490 ±0.010" is when a part is damaged and must be reworked. A skilled technician may never produce a part outside of that range.

These examples expose a substantial weakness in blindly choosing a distribution type without actual manufacturing data, while simultaneously putting 20.000 ±0.500" on the drawing. If the design engineer had used the 20.000 ±0.500" dimension in a tolerance analysis and made an assumption on the distribution (uniform, normal, 6-sigma, parabolic, bi-modal, etc.), that assumption would be wholly invalidated once the technician aimed for a value other than 20.000".

This issue boils down to a potential communication problem. If you must rely on statistical tolerances to produce reliable designs, it is imperative that the tolerances on the drawing drive the manufacturer to statistical machining processes. In other words, if the tolerance analysis assumes that the bar is 20.000 ±0.500" normally distributed (3-sigma Gaussian), the drawing dimension should specify that. In doing so, the technician is not allowed to bias the process, but must produce parts within that normally distributed boundary.

Evaluation Approach

Statistical tolerance analyses are primarily used to help a designer to quantify the dimensional risks associated with a design. These risks can range anywhere from failed builds to catastrophic failures resulting in loss of life. The statistical analysis is there to describe the probability of encountering such a risk, as well as provide other useful information about the variables that most impact the outcome. There are two major approaches to performing a statistical analysis. Each has its benefits and drawbacks. Though they provide strikingly similar information, the reliability of that information may be the subject of contention.

Monte Carlo

Monte Carlo simulations are generally the brute-force approach to performing a tolerance analysis. Let's say we have a tolerance analysis where the length of a rod is a variable. The rod has a length of 4.5 ± 0.5. To perform a monte carlo simulation for using this variable, an actual value (let's say 4.23) for the rod is randomly generated. The value must be between the limits of the dimension; 4.0 and 5.0 in this case. For each variable within the tolerance analysis, a random value is similarly generated. The randomly generated values are summed up (or otherwise combined) to represent a build scenario. This is called a trial. The process is repeated several thousand times in order to produce a better depiction of how the gap would be expected to vary in the real world. Usually performed by a computer, the process of generating random numbers and performing trials can be done in fractions of a second, providing a surprisingly realistic view of what to expect.

The major benefit to using a monte carlo simulation is that interdependent tolerances can be properly modeled and accounted for. Namely, bonus tolerances and datum shifts, which depend on features of size, can be accurately calculated due to the "guessing" nature of the monte carlo simulation. The actual amount of bonus tolerance depends on the as-manufactured size of a feature. The size of the feature is available during a monte carlo simulation and thus lends itself to use in properly calculating bonus tolerance and datum shift.

The major drawback to using a monte carlo simulation is that design iteration and optimization can take a non-trivial amount of time to perform. Because each simulation is a set of trials that cannot stand alone, for each design change, it is likely that a new monte carlo simulation needs to be performed to determine the impact of any changes. Another drawback to a monte carlo simulation is the potential problem inherent to all trial and error processes: Did you perform enough trials? This is only a potential problem because in most cases, we use computers to perform the simulation. And although it becomes a purely academic exercise after about 30,000 trials, we can request more trials to the point where we have no choice but to believe the trend developed by the simulation.

Derivative
Derivative-based simulations are the elegant solution to tolerance analyses. Using partial differential equations to develop a relationship between each variable and the

gap, these types of analyses provide a mathematical model of the gap and its dependency on each variable. For most one-dimensional tolerance analyses, each variable has a one-to-one (or in some cases two-to-one) impact on the gap. That is, for every unit deviation in the variable, the gap deviates by one unit. Determining the impact of a single variable on the gap becomes a simple exercise of multiplying the variable's tolerance range by the partial differential. It answers, very succinctly, the question of how sensitive the results are to a particular variable. And because the model (in most cases) is simply a set of mathematical equations, iterating and optimizing a design and analysis can be done with a few keystrokes.

The major benefit to using a derivative-based tool is that the time to discover impacts of design changes is minimal. It lends itself to fast and effective design optimization.

The major drawback to using a derivative-based tool is its limited ability to represent interdependent tolerances. Where monte carlo tools could properly model and account for bonus tolerances and datum shifts, the derivative-based approach struggles in that the tolerance for a particular feature can depend on the size of the feature. Many software packages attempt to address this potential pitfall by over-compensating for the bonus tolerance and/or datum shift. The result is usually an overly-conservative analysis, though few engineers would complain about this usually-small conservatism. For this reason, it is good to use a one-size-fits-most software package, like QTA, that provides a monte carlo and sensitivity analysis, simultaneously. It also yields RSS results.

RSS

The RSS approach has been around for quite some time. The thought behind the RSS approach is that all tolerances are normally distributed values, so the overall summed impact can be approximated quickly and accurately by hand. Practitioners of this approach typically take the square root of the sum of the squares of all the tolerances.

$$\sqrt{(tol\ 1)^2 + (tol\ 2)^2 + \dots + (tol\ n)^2}$$

The resulting value is taken to represent the upper boundary on the amount of tolerance that will be experienced by the gap 97.3% of the time. This is a very convenient and handy tool to use, but there are some potentially damaging pitfalls to using it.

This is typically only useful and applicable (for tolerance analyses anyway) in one-dimensional tolerance analyses. Of course, there are some very intelligent and mathematically inclined engineers who can properly apply RSS methodology to a two- or three-dimensional problem. Those folks, however, are few and far between. And besides, they are usually smart enough to pick up a 3D tolerance analysis modeling tool instead of wasting a lot of time performing the analysis by hand. The other limitation is that all tolerances should be normally distributed. All warnings concerning improper tolerance distributions apply here. Improperly using an RSS approach can easily result in analysis results that are much rosier than they should be, allowing for the designer to overlook a potentially costly design error.

MRSS

MRSS is a variation of the RSS approach that seeks to account for inherent mean shifts and other manufacturing-related tolerances that invalidate some of the underlying assumptions of an RSS calculation. A couple common variations add scaling factors to add conservatism to the rough calculation.

Bender:

$$MRSS = 1.5 \times RSS = 1.5 \times \sqrt{(tol\ 1)^2 + (tol\ 2)^2 + \ldots + (tol\ n)^2}$$

Gilson:

$$MRSS = \frac{n}{n-1} \times RSS = \frac{n}{n-1} \times \sqrt{(tol\ 1)^2 + (tol\ 2)^2 + \ldots + (tol\ n)^2}$$

Most major companies have a variation of MRSS they prefer over another for varying reasons. They typically take the format of the above equations, however, and generally produce similar results.

FAQ and Solutions

This FAQ and Solutions area is intended to address many of the lingering questions in the "how-to" arena.

Material Modifiers

Q: Can I ignore material modifiers for a quick-look analysis?

A: It is categorically a bad idea to ignore material modifiers in a tolerance analysis. They can singlehandedly make or break a design.

Q: How do I properly deal with material modifiers in a tolerance analysis?

A: Material modifiers allow for a feature to have bonus (or "extra") tolerance. The amount of bonus tolerance depends on the size difference between the manufactured feature size and the virtual size of the feature. The larger the difference in the two values, the more bonus tolerance becomes available to produce an acceptable feature.

For a worst case analysis, you must determine what the worst-case size of each feature is (for some features, it could be MMC and others, LMC), and use the proper tolerance at that material condition in the analysis. This will give you a worst-case view of the analysis because the bonus tolerance will be maximized. For a statistical analysis though, you would want to model the material condition as it is designed to allow for the variation introduced by the material modifier to influence gap as it would in the natural world. In other words, don't pick a material condition for statistical analyses.

Q: How do I properly deal with datum modifiers in a tolerance analysis?

A: Datum modifiers allow for features to have additional tolerance in relation to a datum. Similar to bonus tolerance, the amount of extra tolerance available depends on how much the manufactured size of the datum differs from the virtual size of the datum. Unlike bonus tolerance though, the principal of simultaneous requirements usually allows datum shift within a tolerance analysis to correctly appear only once per feature control frame. Features that have identical feature control frames cannot be impacted by the same datum shift, in opposing directions or magnitudes.

Assembly Slop/Float

Q: How do you account for assembly slop/float in an analysis?

A: There are 3 predominant ways to account for assembly slop. The first two ways are considered superior to the third. No matter how you decide to account for it, there are typically only two variables involved. The size of the male interface (screw shaft, pin, etc.) and the size of the female interface (hole, slot, etc.). The slop is to account for the variation allowed at the interface.

Way #1) Calculate the distance between the outward and inward surfaces of the interfacing features. Basically, this is calculating the amount of air. Let's say you have a screw with a diameter of .112" and a hole that is .122". There is .005" of air on either side of the screw. This method has the practitioner express the air gap as ranging anywhere from .000" to .010". Or .005±.005".

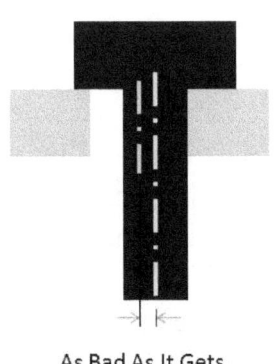

As Bad As It Gets

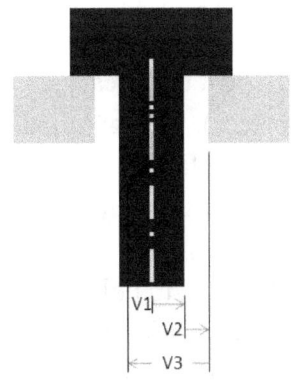

Nominal View with Loop

	C5	▼	f_x =ABS(C6)-ABS(C4)		
	A	B		C	D
1	Gap Description	Assembly slop example			
2					
3	Vector Name	Description		Mean	Tolerance
4	V1	Radius of screw		0.056	
5	V2	Air between screw and hole		0.005	0.005
6	V3	Radius of hole		-0.061	
7					
8					
9					
10		Gap		0	0.005

Figure 31 Assembly Slop Calculation – Way 1

The unique quantity here is V2. The air gap can be as large as .010 or as small as zero. The nominal axis to axis distance is precisely what we would expect: zero. But there can be up to .005 offset between axes.

Way #2) Calculate the distance between the axis or center plane of one feature and the axis or center plane of the other feature. In most cases (if the design is correct), the nominal distance between the axes (or center planes) is zero. But they can be misaligned by as much as the features will allow. This misalignment for a screw is typically expressed as plus or minus the difference in hole and screw diameters, halved. Basically, this allows for the axis of the hole to move to either side of the screw without bias.

$$Axis\ to\ axis\ distance = 0 \pm \frac{Hole\ Diameter - Screw\ Diameter}{2}$$

Way #3) Force the male and female features into contact on one side. This effectively forces the air gap of way #1 to be zero for all cases. It is conservative, but biases a tolerance analysis, rendering it overly-conservative for a statistical analysis. This way is not preferred.

	C5	▼	f_x	=ABS(C6)-ABS(C4)	
	A	B		C	D
1	Gap Description	Assembly slop example			
2					
3	Vector Name	Description		Mean	Tolerance
4	V1	Radius of screw		0.056	
5	V2	Air between screw and hole		0.005	0.005
6	V3	Radius of hole		-0.061	
7					
8					
9					
10		Axis to Axis Distance		-0.005	

Figure 32 Assembly Slop Calculation – Way 3

Either way you choose to account for the assembly slop (whether you use way #1 or 2), the analysis still lends itself to a statistical approach. This is however, another place where a derivative-based statistical approach will not necessarily produce reliable results because the variation of the float is dependent upon the variation of other variables (screw and hole size). For a worst case approach, the features would be assumed to be at their LMC size (biggest hole, smallest screw).

Standard Application

Q: Is assembly slop a positive or negative value?

A: Assembly slop should usually be neither positive nor negative. The sign of the assembly slop describes the direction that the assembly shifts toward. In the case where you must chose, the slop should be taken in the direction that conservatively impacts the gap in consideration.

Assembly Bias Used in Assembler's Favor

Q: Can I use assembly slop in my favor to make an analysis work?

A: It is non-conservative to assume that assembly slop can be used in the assembler's favor during a build. In many cases, there is no slop available to take advantage of. Consider the case where there are multiple holes and screws attaching two components to one another. The availability of assembly slop depends largely on the position of the pattern of holes and screws. If two holes are spread apart as much as the tolerances allow, and the corresponding screws are spread as much as tolerances allow, this could easily present a scenario where the parts are not allowed to float relative to one another whatsoever, in any direction. If you are in a position where you are relying on the availability of assembly slop to permit assembly, it is best to approach it with caution. Be sure that the assembly float you wish to take advantage of is at a stage in the build process that the fasteners are being installed and the gap in question is available for inspection. For the analysis, it is more realistic to assume that the hole pattern can be centered on the screw pattern rather than biasing the entire subassembly in a particular direction. In other words, make the assembly slop zero in your calculations. It is much easier to assume that a part is centered up than to try to keep track of which way to bias a part during a build. Remember – you can only bias a part in one direction at a time.

Multiple Assembly Constraints

Q: How do I deal with multiple assembly constraints?

A: Parts that have multiple and potentially competing assembly constraints cannot always be avoided. It is best to perform the tolerance analysis using different constraint paths, and choose the path that produces the result with the least variation. This is the most constraining assembly path.

Fastener Equations

Q: What is the formula for a fixed-fastener configuration?

A: A fastener is considered "fixed" if the screw is not free to float relative to one of the two parts being clamped together. This is possibly the most common fastened type in most production environments. The following equation is used to justify the thru-hole size and positional tolerance and the positional tolerance of the female threads within the mating part:

$$H - T_1 \geq F + T_2 \text{, where}$$

H is the hole size at maximum material condition, T_1 is the positional tolerance of the hole at maximum material condition,

F is the size of the screw, and T_2 is the positional tolerance of the threads within the mating part.

It is important to note that this equation assumes that a projected tolerance zone is specified on the female threads.

Q: What is the formula for a floating-fastener configuration?

A: A fastener is considered "floating" if the screw is free to float relative to both parts being clamped together. The following equation is used to justify the thru-hole sizes and positional tolerances within each part:

$$H_1 + H_2 - T_1 - T_2 \geq 2F \text{, where}$$

H_1 is the hole size within Part 1 at maximum material condition,

T_1 is the positional tolerance of the hole within Part 1 at maximum material condition,

H_2 is the hole size within Part 2 at maximum material condition,

T_2 is the positional tolerance of the hole within Part 2 at maximum material condition, and

F is the size of the screw.

Double-Fixed Fasteners

Q: How do you analyze double-fixed fasteners?

A: Double-fixed fasteners should be avoided whenever possible. Because it is not always possible, it is necessary to perform a tolerance analysis to determine that mounting hole sizes and positional tolerances are sufficient to promote proper fit. Though floating and fixed fastener conditions have simple equations that can be used for evaluation, there is no equation to evaluate a double-fixed fastener. In other words, a tolerance loop is required to validate the joint.

Assembly Biases

Q: Under what circumstances should I assume that an assembly bias exists?

A: Assembly bias is usually present when a force is present during assembly that is consistent in its presence and direction. Gravity is, by far, the most common assembly biasing agent, but there are other biasing agents. Among them are springs, ramped surfaces, and torque operations.

Form Tolerances

Q: When should I consider form tolerances in a tolerance analysis?

A: There are many instances where form tolerances, such as flatness and circularity, should be considered in a tolerance analysis. The most common instances are when one of the following conditions exists:

1) The mounting surface of one part interfaces to a small portion of the mounting surface on another part.

2) The analysis is evaluating wall or edge thicknesses.

3) The design documentation waives the perfect form at MMC requirement on a feature or surface within the analysis.

More Tolerance Improving Yield

Q: My tolerance analysis is showing that my yield increases when my tolerances increase. Is something wrong?

A: When this happens, it is usually indicative of a design problem. The nominal values of the design are likely yielding an unacceptable design case. In other words, the design is relying on the manufacturer producing parts and features that are far from nominal in order to meet the design requirements.

Q: How do you deal with thermal expansion?

A: In the vast majority of cases, the magnitude of any dimensional changes experienced by thermal expansion is minimal and negligible. In the rare instances where thermal expansion must be considered, the difference between the coefficients of thermal expansion (CTE) of individual parts is most critical. Care must be taken to adjust (increasing or decreasing) individual dimensions, based on the temperature that the analysis is taking place at. Typically, dimensions must meet drawing requirements at 68°F, but could, in theory, go outside of the allowable tolerance range at temperature extremes.